Earthquakes and their Causes

Earthquakes and their Causes

LM Publishers

Preface

Earthquake is the result of a sudden release of energy accumulated in the earth's crust usually caused by movement along a fault plane or by volcanic activity. The result of the breakdown of surface rocks is called a fault. The point of the initial rupture is called focus or hypocenter, and the location directly above it on the surface of the earth is called the *epicenter*.

It is estimated that there are 500,000 earthquakes in the world each year, detectable with current instrumentation. About 100,000 of these can be felt.

The effects of earthquakes include the following : shaking and ground rupture, landslip and avalanches, fires, soil liquefaction, tsunami (we remember the recent tsunami of 2004 in asian countries), floods and human impacts.

Scientists developed many methods for predicting the time and place in which earthquakes will occur. Despite their considerable research efforts, scientifically reproducible predictions cannot yet be made to a specific day or month. The following works help us to know more about the earthquake phenomena and their causes.

The causes of Earthquakes[1]

The causes of earthquakes have long been the subject of many conjectures. The numerous investigations during many decades have contributed much to define their characters; and several data recently acquired tend further to make their mechanism clear. It is known that the shocks are by no means distributed at haphazard over the surface of the globe. The countries where the strata have preserved their original horizontal position, like the north of France, a part of Belgium, and the most of Russia, are privileged with tranquility. Violent commotions are manifested particularly in regions that have suffered considerable mechanical accidents, and have acquired their last relief at a recent epoch, like the region of the Alps, Italy, and Sicily.

---

[1] By M. Daubrée

The tracts that are simultaneously disturbed by the same shock most frequently comprise arcs of from 5° to 15°, or from 300 to 1,500 kilometres. They rarely include a much more considerable fraction of the globe; although the celebrated catastrophe at Lisbon on the 1st of November, 1755, extended over some 17° or 18°, into Africa and the two Americas, or over a surface equal to about four times that of Europe.

The detailed examination of many earthquakes has enabled us to determine the center of the shocks as well as the contours of the disturbed areas. From the manner in which the latter surfaces agree with the lines of pre-existing dislocations, several of the most distinguished geologists, including Mr. Dana, M. Suess, and Albert Heim, have considered the shocks in question as connected with the formation of chains of mountains, of which they may be a kind of continuation.

In fact, the crust of the earth everywhere shows the enormous effects exercised by the lateral pressures that have been in operation at all epochs. The strata, bent and bent over again many times through thousands of metres of thickness, as well as the great fractures that traverse them, are the eloquent witnesses of these mechanical actions. Notwithstanding the apparent tranquility now reigning on the surface of the globe, equilibrium does not exist in the earth, and commotions have not been arrested in its depths. The proof of this is found, not only in earthquakes, but also in the slow movements of the soil, of elevation and depression—a kind of warping, which has continued to manifest itself within historical times in all parts of the globe. It is conceivable that slow actions of this kind, after more or less prolonged strains, may end in abrupt movements, as Elie de Beaumont supposed. We can see, also, in experiments intended to imitate the bending of strata, how gradual inflections lead all at once to fractures and outbursts. Simple cavings in deep cavities, have also been regarded as possibly giving rise to earthquakes; and this opinion has been adopted by M. Boussingault after the well-known observations he made in the Andes. There is, in fact, nothing to prove that disturbances of these different kinds do not take place in the interior of the globe; but we may certainly consider them as the general cause of earthquakes. These shocks are, however, most commonly in evident connection with volcanoes; and it is in the neighborhood of the latter that they are especially frequent. As is well known, every volcanic eruption is announced by precursory earthquakes, the violence of which is stilled when an outlet is opened for the vapor of water which is successively the cause of the subterranean agitations and the projecting agent of all the eruptions. The tension of the vapor in the volcanic reservoirs must be very high. Thus, that pressure which forces the lava up to more than 3,000 metres above the sea, to the top of Etna, cannot be less than a thousand atmospheres.

An attentive study of the phenomena confirms the attribution of the cause of the shocks, however violent they may be, to the vapor of water. It is sufficient for this to be the case for vaporization to take place at a temperature of 1,000° C. (1,800° Fahr.), approximately that of lava, and under a volume equivalent to that of the water in the liquid state whence the vapor is derived. Under these conditions, we must suppose the vaporization to be total, for the critical temperature, above which the liquefaction of vapor cannot be realized, is, according to M. Clausius, 332° C. (629° F.). The pressure, of which it is also possible to make an approximate estimate, then becomes comparable to that of the most powerfully explosive gases, and is, consequently, capable of producing very considerable dynamic effects. These effects may also be produced at a much lower temperature than that of lavas at 500° C. (900° F.); for example, if we suppose that the volume imposed upon the vapor is so limited as to correspond to a density of 0·8 or 0·9. No doubt such conditions are realized in the lower regions of the globe, where water is confined within limited spaces, and as hot as the melted rocks which we see gushing out from the surface at a temperature of 1,000° C. (1,800° F.) or more. We shall see, however, that such depths and such a temperature are not necessary.

The vapor of water when superheated acquires a power of which the most terrible boiler-explosions could give no idea if we had not the result before our eyes. The tubes of the best quality of iron that I used in observing the action of superheated water in the formation of silicates had an inside diameter of twenty-one millimetres and were eleven millimetres thick. They sometimes exploded, and were projected into the air with a noise like that of the firing of a cannon. Before bursting, the tubes swelled out into bulbous forms, and rents were opened in the middle of the bulbs. If the iron had no flaws and according to the estimate that it would preserve up to 450° C. (810° F.), the temperature to which it was raised, the same tenacity it had when cold, such rents must have indicated a pressure of several thousand atmospheres. A few cubic centimetres of water were sufficient to produce an effect like that; and, considering the small dimensions of the inside of the tubes as compared with the volume of the water, the vapor must have reached a density of about 0·9. If we apply the data we possess to the depths of the globe, it is not difficult to conceive very simple dispositions in which the vapor of water, under the conditions we have just determined, will suddenly provoke shocks or series of shocks that will too often make themselves felt on the surface. Whatever conception we may form of the volcanic reservoirs, we must admit it to be very probable that solutions of continuity exist between the soft or fluid masses in fusion and the solid masses superposed over them. Moreover, cavities may also exist in the solid rocks themselves that lie over the soft masses. On the other hand, the incessant losses, which these internal reservoirs suffer in consequence of the enormous volumes of water in the condition of vapor which they disengage every day, are probably repaired by supplies from the surface.

I have shown by experiment that these supplies may be transmitted through the pores of some kinds of rocks. Simple capillary action, in conjunction with gravity, may force water to penetrate against very strong counter-pressure, from the superficial and cooler regions of the globe, to deep and hot regions, where, by reason of the temperature and pressure it acquires there, it becomes capable of producing very great mechanical and chemical effects. If we suppose that water penetrates, either directly or after a halt in a reservoir where it has remained liquid, to masses in fusion, so as to acquire there an enormous tension and an explosive force, we shall have the cause of the anterior real explosions and of the instantaneous shocks due to gases at high pressure. If the cavities, instead of forming a single reservoir, are divided into several parts or distinct compartments, there is no reason why the tension of the vapor should be the same in the different receivers, provided they are separated by walls of rock. The pressure may even be very different in two or more of them. This admitted, if a separating wall is broken by excess of pressure or melted by the heat, vapor at high pressure will be set in motion, and in the presence of the solid masses upon which it will strike it will behave just as if there had been an instantaneous formation of vapor, as we supposed in the former case.

It is very hard to establish, as has been attempted, a clear line of demarcation between the character of the earthquakes of volcanic regions proper and of regions without volcanoes, such as Portugal, Asia Minor (Chios, April 3, 1881, five thousand victims), Syria, Algeria, and the rim of the Mediterranean generally. In both classes, the characteristic manifestations which we perceive are the same. If, as some assume, the internal movements of the rocks were a cause of real earthquakes, it could only be because those internal movements mechanically developed heat, and in that way provoked the formation of vapor. But, in the recently disturbed regions we have especially in view, which are the seat of so frequent shocks, another cause is much more probable. There doubtless remain in them interstices and interior cavities that permit the access of water to the hot regions. The depth of the centers of disturbance of earthquakes has been estimated, in different cases, by calculations only grossly approximate, at eleven kilometres, twenty-seven kilometres, and thirty-eight kilometres. In any case, such depth, though very slight in comparison with the length of the radius of the earth, is great enough for the temperature at the normal rate of increase to be very high; and the same will also be the case with the water that may be present there. Now, as we have already seen, a temperature of 500° C. (900° Fahr.) is sufficient to cause water to explode with violence.

It is certainly in the largest number of cases very difficult to admit collisions of solid bodies in the interior as the moving causes of earthquakes. How, for instance, can we conceive that so violent and extensive an earthquake as that of Lisbon on the 1st of November, 1755, was produced in this way? John Mitchell (Royal Society, 1760, vol. x, p. 751) drew from this memorable example the conclusion that the vapor of water intervenes in these shocks as well as in the eruptions of volcanoes. Manifest effects of a class of internal explosions, undoubtedly due to the production or sudden moving of a great quantity of superheated vapor, are exhibited at the present epoch, and are not rare. Such explosions, for instance, are exceptionally formidable in the region of Java, and the mind is naturally led to the one which has just convulsed the zone between that island and Sumatra, which has caused the disappearance of the island of Krakatoa and its mountains, has raised other mountains, and has claimed more than forty thousand victims.

At a period more remote from us, the explosive force of interior gases gave rise to very remarkable circular cavities, which have been called "craters of explosion", and are well known. Examples of them are found in Auvergne (Lake Pavin) and in the district of the Eifel, where the stratified beds have been sharply cut as if with a punch. What gases thus put in motion are capable of, as a mechanical power, could hardly have been suspected till since the explosive effects of gun-cotton, nitroglycerine, and dynamite, have been known. The effects of compressed air in the air-gun and of the powder-gases in fire-arms have been wonderfully surpassed, for we now measure explosive pressures of six thousand atmospheres and more. In the experiments in which I have had occasion to observe gases at high pressure in order to explain the action that a meteor coming with planetary speed is subjected to on the part of the atmosphere into which it plunges, I have been surprised at witnessing the great energy of gaseous masses. They engrave themselves deeply, as if with a burner, into the pieces of steel that are opposed to them, and of themselves reduce a part of it to an impalpable dust shot into the atmosphere as if it were volcanic ashes. It is no less surprising—and this observation is of much importance in explaining the problem that occupies us—to remark the tenuity of the gaseous mass that produces such results. Yet its force causes ruptures which the pressure of a weight six hundred thousand times heavier than the gas could not effect!

In short, gaseous movements under high pressure, put in operation from time to time by a simple mechanism like what Nature can and does present, will account for all the essential features of earthquakes. Much better than the hypothesis of interior collisions of solid bodies, they explain the effect of the shock, resembling the blows of a ram, their violence, their frequent succession, and their recurrence in the same regions after many centuries; they explain also the production of earthquakes in regions of dislocation, especially in those in which the disturbance is recent, and their subordination to deep fractures of the crust of the earth.

Earthquakes seem to be volcanic eruptions that are suppressed because they cannot find any outlet, nearly as Dolomieu thought. The motive power of gases, the immense effects of which we can see in the protuberances or jets shot out from the sun with prodigious speed and of enormous dimensions, appears to be sufficiently considerable in the depths of the globe also to explain all the effects of earthquakes.

The origin of Earthquakes[2]

The origin of earthquakes has been assigned to many causes, as the falling in of caverns, steam, the combustion of gases, volcanic and electric action. Their most prominent and peculiar features are the following :

1. Great subterranean noises and reports resembling thunder. These occur more or less during all earthquakes. Father Kircher describes them as "a horrid sound resembling that of an infinite number of chariots driven fiercely forward, the wheels rattling, and the thongs of the whips cracking;" Sir Hans Sloane, in Jamaica, as "a hollow rumbling noise almost like that of thunder." At Colares, near Lisbon, in 1755, during the great earthquake, the sound is said to have been "like that of chariots, which increased till it equaled that of the roar of

---

[2] By Lake J. John

cannon;" and at Lisbon, "a rattling as of coaches in the street, with a frightful noise underground resembling the rumbling of distant thunder." At Madeira the same earthquake was preceded "by rumbling noises in the air like that of empty carriages, which died away like a peal of distant thunder," On the 16th of September, 1849, there was an earthquake at Burra-Burra, in South Australia, where the noise is said to have resembled the rolling of heavy carriages. The shock was followed by a flash of lightning that illumined the whole atmosphere.

2. Another feature of these phenomena is the upheaval of the ground observed during the prevalence of most earthquakes, which is one cause of the sea retiring, another being the suction of the approaching wave when the centre of the convulsion has been removed from the shore. During the great earthquake at Lisbon the bar at the mouth of the Tagus was laid bare by the upheaval, and the master of a vessel, lying in that river at the time, stated that his large anchor was thrown up from the bottom, and seemed to swim on the surface of the water. Other results of the upward

movement during this catastrophe were observed elsewhere. The water in a pond at Dunstal, in Suffolk, was jerked up into the form of a pyramid. At some places the water was tossed out of the wells. At Loch Lomond a large stone was forced out of the water. Rocks were raised into the air from the bottom of the Atlantic, and on board a vessel, about forty leagues from the island of St. Vincent in the West Indies, the anchors, which were lashed, bounced up, and the sailors thrown a foot and a half perpendicular from the deck, the ship sinking into the water immediately afterward as low as the main-chains. At Riobamba, in South America, on the 5th of January, 1797, the bodies of many of the inhabitants were thrown, by this vertical action, upon the hill of La Cullca, which is several hundred feet high, and on the opposite side of the river. During some of these convulsions in Italy, paving-stones have been tossed into the air and found with their lower sides uppermost; and, at the time of a late convulsion in South America, the rising of the ground caused the sea to retire, which returned like a wall in appearance, carrying before it inland vessels that had only a few minutes before been left dry, towns and people being overwhelmed by the resistless recoil.

3. Another peculiarity to be noticed in these convulsions is the frequent horizontal and circular motion of the soil. These effects are often very curious, and, in countries much subject to such catastrophes in their severest forms, have often given rise to lawsuits. Walls that had served to divide fields have been completely changed in direction, but without having been shattered or overthrown. Straight and parallel rows of trees have been inflected, and fields and portions of fields have changed places. Houses have also exchanged situations with each other.

4. It has been observed that clouds have become fixed or suspended over particular spots affected, or about to be affected, by earthquake, as in London, in 1749, in Calabria, in 1783; and it is more than probable that the fog that enveloped Euphemia, in Sicily, in 1638, Millitello in 1693, and other places when they were destroyed, arose from the operation of one cause.

5. Explosions of great violence frequently attend these convulsions, often with disastrous results. When Millitello was destroyed, there was a great explosion heard in the fog that enveloped it; traces were noticed afterward as of the presence of fire on the rocks in the neighborhood, and the vines in the country surrounding it appeared as though they had been seared by fire. A similar explosion was heard in 1783 at Castel Nuovo, in Calabria, when that place was overwhelmed.

6. A further peculiarity is the exemption of certain spots, although the shocks were felt in all the surrounding neighborhood. Thus, at Manchester, in 1777, St. Paul's Church and the Dissenting Chapel escaped. Both of these were low buildings without steeples, and the church situated over a common sewer; but other more lofty buildings, especially those with metal pipes attached, felt the shocks severely. At Blockley the shocks were experienced strongly at the church, but very slightly at the chapel about 300 yards distant, and the latter was constructed without water-pipes.

7. Earthquakes are very frequently attended by thunder and lightning. At Munster, in 1612, thunder and lightning were heavy during an earthquake; and in Sicily, in 1693, it caused very great mischief. This conjunction of lightning with earthquake was noticed by Luke Howard, and constitutes what he designates "spurious earthquake." One of the cases he mentions occured in Radnorshire: "At Knill Court the oscillation of the house was plainly perceptible, and felt by all the family, and that, too, in several apartments, and was accompanied by a peculiar rumbling noise. At Harpton, a severe storm of thunder and lightning was experienced the same night and at the same time."

8. Peculiar rushing noises have also at times been perceived, as in Staffordshire in 1692, and London in 1749.

9. These convulsions are attended by the disturbance of the magnetic needle, and compasses on board ship are frequently for a time useless. On the 19th of January, 1845, on the Thames steamer, during an earthquake in the West Indies, they revolved on their pivots

with great rapidity; and on the 29th of October, 1867, during a hurricane, there were shocks of earthquake at St. Thomas's, and the electrical disturbance was so great as temporarily to render the compasses unavailable.

Such being some of the more prominent peculiarities attending earthquakes, let us now apply them to the theories above referred to, and endeavor to ascertain the causes of these disturbances or the agencies employed in producing them.

They do not support the theory of the foiling in of caverns being the cause of these phenomena; for they are invariably attended by an upheaval of the ground, and often with a horizontal or a circular motion. This theory, therefore, cannot be maintained, and more especially as it does not explain the electric and magnetic accompaniments.

The hypothesis that they are caused by steam or the explosion of confined gases has scarcely a better foundation. These agents might produce vertical motion and subterranean noises, but it is difficult to conceive how they could bring about circular motion at the surface; and it is quite impossible that the explosion of gases or the escape of steam could, immediately preceding a shock, attract the clouds floating in the atmosphere, so that they should remain fixed over particular spots. Other characteristics also cannot' be explained on this theory, as the lightning and disturbance of the compass.

The volcanic and igneous theory is not so easily to be disposed of; for it appears very clear that volcanic eruptions do produce carthquake. A remarkable instance is that of Santorini in 1650. Earthquake is also very common where volcanic action is extensively developed, as in South America and the neighborhood of Etna and Vesuvius.

Volcanoes produce these disturbances in two ways: 1. By their own direct motion; 2. By disturbing the electric equilibrium in their neighborhood. This electric disturbance was noticed by Pliny, who records that an officer, one of the Decuriones Municipales of Pompeii, was struck by lightning in 79, although the sky was perfectly unclouded; and these indications have been put to practical use. The presence of lightning is also a prominent feature during volcanic eruptions. When Kattleguia, in Iceland, now extinct, was last in a state of eruption, lightning proceeded from it and killed a farmer and his servant, together with some horses and cows. We cannot, therefore, exclude the consideration of volcanoes as producers of earthquake, sometimes by direct action, at others through the medium of electric disturbance.

But by far the most prominent agent seems to be electricity, and the Italians, who suffer so much from these calamities, consider it to be the only cause. The evidences of the activity of the electric fluid in this respect are so palpable that they cannot be controverted. As some may be skeptical on this point, it will not be amiss to examine a few cases in which the operation of this agent is quite apparent.

When considering this part of the subject, we must not omit to notice the frequency with which the greatest weight of these calamities falls upon towns and the neighborhoods of mountains. This is to be accounted for on the electric theory, from these places offering points for the escape of the fluid which naturally flies there to seek a thoroughfare, so to speak. From this cause we have St. Elmo's fire on the masts and yards of ships at sea, and De Saussure's experiences of the escape of the fluid from an Alpine peak. Hence we may infer that towns and mountains create centres of force in these convulsions.

At Münster, in Germany, an earthquake began on December 8, 1612, and lasted for several days. During the shocks, Billenelt Castle, near Münster, built on a rock, "sunk more than the depth of two men's height," a breach being made in the rock itself. The destruction by earthquake and lightning seems to have been great. "If any," says a chronicler of the catastrophe, "have so much heart left as to lift up his hands to heaven, he is presently struck down by thunder and lightning;" "fiery clouds and a direful comet" alarmed the superstitious. The state of the atmosphere must have been very peculiar, even allowing for exaggeration, since the writer referred to states that the appearance of the stars was "changed into prodigious, dreadful, fiery meteors."

During this calamity, earthquake, thunder, and lightning, occurred twice every day, but not at the same time.

The earthquake of 1638 disturbed both Etna and Stromboli, causing them to send forth flame and smoke, as though the sources of the convulsion descended deeper than their roots. Father Kircher describes the disappearance of the city of Euphemia, which he was endeavoring to reach at this time, and was in sight of. After a violent shock, on rising from the ground and looking toward the city, he saw only a frightful dark cloud, which surprised him and his companions, as the sky was otherwise very serene. Waiting until the cloud had passed away, they found Euphemia had totally disappeared, and its place a putrid lake.

The earthquakes of 1692, in Jamaica, and 1693, in Sicily, present very strong evidences of general electric disturbance in the globe at those times. One evening in February, 1692, at Alari, in Sicily, the village seemed to the country-people to be in flames. The fire, as they imagined, began by little and increased for about a quarter of an hour, when all the houses in the place appeared to be enveloped in one flame which lasted about six minutes and then began to decay, as from want of more fuel. Many who ran to render assistance observed this increase as they passed along the road, but on entering the village found all to be a delusion. Such appearances of fire and light occur in other localities subject to earthquake, e. g., at Cowrie, Perthshire, one morning before daybreak, in 1842, the light is stated to have been so brilliant that birds were distinguished on the trees. Again, in Sicily, about the 15th of May, following the incident at Alari, two hours before sunset, the atmosphere being very clear, the heavens appeared on a sudden all on fire, without any flashes of lightning or the least noise of thunder. This lasted, at Syracuse, about a quarter of an hour, when there appeared in the air over the city two bows, the colors extremely bright, after the usual manner, and a third with the extremities inverted, and, as not a single cloud was visible in any part of the sky, the abnormal state of the atmosphere is clear. It was also during this summer that the unusually severe thunder-storm occurred at Geneva that so materially affected the future career of the celebrated Robert Boyle. The earthquakes at Jamaica began on the 17th of June, and their greatest violence seems to have been spent in the mountains. Terrific noises were heard among them at Port Royal during the last shock, and they were so torn and rent as to present a very shattered appearance and quite new forms. In this month Etna emitted extraordinarily loud noises for three days together. A singular circumstance, during this catastrophe at Jamaica, was the derangement of the wind. The land-breeze often failed, and the sea-breeze blew all night, whereas the land-breeze should blow all night and the sea-breeze all day. There was an earthquake on September 8, 1692, in Europe, but I have not yet been able to find out the locality.

Space will not admit of more than noticing some special phenomena of the Sicilian earthquakes, 1693. On the 10th of January the castle of Augusta was blown up by the lightning firing the powder-magazine. At Minco, on the 11th, the shock was attended by "a mighty storm of lightning, thunder, and hail, that lasted six hours." The archbishop's palace at Monreal was set on fire by the lightning. Etna emitted great noises, flames, and ashes, during the shocks that overthrew Catania, but there does not appear to have been eruption. Furla, situated among limestone-quarries, disappeared, and at several parts of the hill the rocks, which were previously almost as white as Geneva marble, had changed, and in the clefts made by the earthquake had become of a burnt color, as if fire and powder had been employed to rend them asunder. Millitello seems to have been destroyed before the 11th of January, for the country-people, who dwelt on the neighboring ridge of mountains, affirmed that it was not to be seen on the morning of that day, to which time, from twelve o'clock on the 8th, it had been concealed in a thick fog. During the interval the mountain that lay on the north side of the town had been split asunder— one portion overwhelming Millitello, so that not an inhabitant escaped. Francofonte, built chiefly of wood, escaped with little damage from the shocks, but was fired by lightning; the spire of the church—wood covered with lead—

burnt down, and the nunnery of the Carmelites entirely destroyed so suddenly, that five of the nuns were stifled in their beds. The largest part of the inhabitants of Luochela escaped by flying from the town on the sudden disappearance of the castle, situated on a rising ground. Ragusa experienced shocks on the 8th, with violent thunder and lightning. At Specufurno, on the 10th, "from morning till night, there was never heard so violent a storm of thunder and lightning, as if heaven and earth had been mixing together;" the town-house and several other houses were destroyed by it. The peasants on the neighboring hills observed that this lightning had burnt the vines so that no crop could be expected for the season.

The earthquake of London, 1749, also exhibited strong symptoms of electric action. The year abounded with thunder and lightning, coruscations frequently appeared in the air, and the aurora removed to the south, showing upon two occasions unusual colors. Dr. Stephen Hales heard a rushing in his house which ended in an explosion in the air as from a small cannon, and attributed it to the escape of the fluid by the steeple of the church of St. Martin's-in-the-Fields, adjoining. The Rev. J, H. Murray refers to the electrical disturbances on the east coast of South America, contemporaneous with the great earthquakes on the west coast in 1868, and considers them related. He describes one storm, just at the time of the earthquake, as giving "an idea of what the bombardment of Sevastopol must have been like."

The phenomena of seaquake are of a similar character. We have ourselves seen electric clouds thrown into auroral forms contemporaneously with the disturbance of the sea at another locality.

Examples might be extensively multiplied, but the above would seem sufficient to show that a leading cause of earthquake is electric action, and that volcanoes sometimes produce the same by direct convulsion, and at others by disturbing the electric equilibrium of a locality.

Earthquake Phenomena[3]

In the afternoon of the 1st day of June, 1638, 18 years after the landing of the pilgrims, there occurred the first earthquake in New England, of which we have an authentic record.

It is 234 years since that event, and, according to a catalogue prepared by W. T. Brigham, published in the Memoirs of the Boston Society of Natural History, it appears that, down to October 20, 1870, 231 earthquakes are recorded as having taken place in New England. From this able paper we learn that, in 1663, portions of Canada, New England, and New York, were convulsed by earthquake-shocks.

In 1727, at Newbury, and near the mouth of the Merrimac River, an earthquake took place during the evening when the atmosphere was perfectly calm and clear. The sound preceded the shock. The earth opened, and a

---

[3] By Elias Lewis

sulphurous blast threw up mounds of calcined dust. Several days previous to the earthquake, water in the wells became fetid, and of a pale brimstone color. In 1755, on the 18th day of November, a hollow, roaring noise was heard in various parts of New England. In a minute the earth seemed to undulate as if a wave were passing. This was followed by a vibratory and jerking motion, familiar in earthquake countries. The first shock of this earthquake occurred 17 days after the terrible one at Lisbon, the vibrations of which had not yet ceased.

The great earthquake at New Madrid, in Missouri, took place in 1811-'12. The shocks here were vertical, proving, as we shall see hereafter, that the centre of energy was directly underneath. At other times, the shocks, which continued many months, were undulatory. The ground rose in huge waves, which burst, and volumes of water, sand, and pit-coal, were thrown high as the tops of the trees. The forests waved like standing corn in a gale of wind, and an area 70 miles long by 30 miles wide was submerged, and became a swampy lake.

On the 13th of August, 1868, a fearful earthquake took place in Peru, which laid waste much of the country lying between the Andes and the Pacific. The shocks were felt through a distance of 1,400 miles north and south, and three important cities were destroyed. At Arequipa, in Peru, 40 miles from the sea, a slight undulatory shock was felt, followed by others so violent that in five minutes not a house was standing in that city of 44,000 inhabitants. A subterranean rumbling, like the rush of an avalanche, was heard above the crash, and a cloud of dust rose in the still air over the city. On the sea-coast were situated Iquique and Arica—both were destroyed by the shocks, and overwhelmed by a tremendous wave. The ocean thus took up the vibrations of the land, and waves of tremendous volume were put in motion, which rolled, not only upon the coast, but away from it with a velocity in the deep ocean of not less than 400 miles an hour. The great wave—for one was of much greater volume than the others—has been estimated at upward of 200 miles breadth, with a length along its curved crest of 8,000 miles.

This rolled into the harbor of Yokohama, in Japan, 10,500 miles distant, and was felt at Port Fairy, in South Victoria, distant nearly one-half of the earth's circumference.

In 1797, a province of Ecuador, about 100 miles south of Quito, was visited by what is described by Humboldt as "one of the most fearful phenomena recorded in the physical history of our planet." The shocks were vertical, and occurred as "mine-like explosions." The town of Riobamba was over the central area, and many of its inhabitants were thrown 100 feet into the air.

The shocks, in this instance, were not announced by any subterranean thunder, but, just 18 minutes after, a terrific roar was heard beneath Quito. It thus appears that shocks are not always preceded by sounds, nor do the sounds increase with the violence of the shock.

Sometimes, says Humboldt, there is a "ringing noise, as if vitrified masses were struck; again, a continuous rumbling and hollow roar; at others, a rattling and clanking as of chains or near thunder." With the lightning's flash we know that the danger is over, and await the coming thunder without alarm; but thunder, rolling deep in the earth, announces possible if not certain calamity.

Throughout the region of the Andes a connection between volcanic and earthquake action has been recognized by the people. It was supposed by Strabo that volcanoes are safety-valves, and scientific observation suggests that they may relieve the pressure and tension which would otherwise lay the earth in ruins.

For two years previous to 1538 earthquakes had been violent and frequent at Pozzuoli on the Bay of Baise, and elsewhere in the vicinity of Naples. On the 27th and 28th of September they did not cease day or night. On the night of the 29th, flames issued from the ground near the baths of Nero, the earth rose and burst at the top with tremendous roar, and discharged steam, gas, pumice, mud, and ashes. A mountain 1,000 feet high was formed, known as Monte Nuovo, which, at the present time, is 8,000 feet in circumference, and 440 feet above the bay.

The Phlegræan Fields, of which Monte Nuovo now forms a part, have, in the opinion of Sir Charles Lyell, a "mutual relation with Vesuvius—a violent disturbance in one district serving as a safety-valve to the other—both never being in full activity at once."

In the Sandwich Islands, in 1868, Mauna Loa and the craters of Kilhauea on its flank were in active eruption. The valleys of the mountains were filled with rivers of fire, and a cloud of smoke and vapor arose, it is said, over the mountain, to a height of eight miles. During these fearful phenomena, which continued more than a month, 1,500 earthquake-shocks occurred, 300 of which were counted in five days. But whether shocks occur in the immediate vicinity of volcanoes during eruptions, or whether activity of the one diminishes the violence of the other, it is certain that they have a mutual relation, and probably a common origin.

The opening and closing of fissures and chasms in the ground during earthquakes is a common phenomenon. Men, animals, and dwellings, are sometimes swallowed in them, and forever disappear. In 1848 an earthquake shook a large portion of New Zealand, and a fissure of great depth opened along a chain of mountains from 1,000 to 4,000 feet high, extending 60 miles, but of not more than 18 inches in average width.

During the Calabrian earthquake of 1783 the surface of the ground opened and closed in immense fissures, by means of which new lakes were formed and others drained or were dried up.

At Jerocarne the earth is described, by Sir Charles Lyell, as lacerated in an extraordinary manner. "Fissures ran in every direction, like cracks in a broken pane of glass."

In another instance, several dwellings were engulfed in a fissure, and were found to be jammed with their contents into a compact mass. Chasms of immense length and depth were formed. Some were crescent-shaped, and a mile in length.

The plains of Calabria were covered in many places with circular hollows from one foot to three or four feet in diameter. Some of these were filled with water, others with dry sand.

But changes in the earth's crust occur during earthquakes, on a still grander scale. Evidences of local disturbance, however disastrous it may have been, are often effaced if not forgotten in a few centuries frequently in a few years. But the slow upheaval of mountain-chains and the dislocation of strata through profound depths are results which alter at last the physical aspect and contour of the surface of the globe. It would not be proper, however, to say that these changes are caused by earthquakes, but rather that the earthquake vibration is a concomitant of the displacement by which they are produced.

Humboldt, Lyell, Dana, and other authorities, consider earth quakes to be the dynamic result of action of the earth's heated interior upon its cooled exterior. Whether the central portions of the earth be fluid or not, it is quite certain that heat increases as we descend; and it is estimated by Sir Charles Lyell that the heat at a depth of 25 miles would be sufficient to melt granite, and at 34 miles to render fluid or incandescent every known substance. We have no means of knowing the condition of matter under the enormous pressure which prevails at a depth of 34 miles, and are most concerned with the fact that the heat of fusion exists at no very great depth beneath the surface. The earth's crust is, therefore, its cooled exterior.

It is found that nearly all rocks contract by cooling and expand by heat. Lyell estimates that sandstone a mile in thickness, and heated to 200° Fahr., would expand so as to lift a mass of rock upon it 10 feet above its former level; and if a mass of the earth's crust equally expansible, 50 miles in thickness, be heated to 800°, it would rise 1,500 feet. From cooling we have the reverse effect—shrinkage, contraction, lateral pressure, and ultimately bending of the strata.

The strain thus produced will at last cause fracture, and the vibration that results is an earthquake.

This form of tension is being continually and everywhere produced in the earth's crust, and there is probably no instant of time when that crust is entirely free from vibrations.

"There is nothing," observes Darwin, "not even the wind that blows, so unstable as the level of the crust of the globe."

Prof. Tyndall observes that, "where the acting force is small and the time great, the result is a slow and almost inappreciable change." Thus, great areas of land may be elevated or depressed. "But where the intensity is great and the time small, sudden convulsion must ensue." Thus, in an instant, mountains may undergo a change of elevation, or be shaken to fragments, or tracks of land sunken and over-flowed. In the delta of the Indus are extensive areas of level ground, over which native villages were scattered, with fortifications and other defences.

In 1822, just half a century ago, an earthquake occurred in Chili, of terrific violence, even for that region of convulsions. It was estimated that 100,000 square miles of land were elevated from two to seven feet, the rise being greatest inland, and probably included a portion of the Chilian Andes. The location of the force must have been at great depth, perhaps not less than 20 miles below the base of the Andes; and it is probable that the entire superincumbent mass underwent a change of level of from two to seven feet of perpendicular elevation.

The earthquake at Lisbon, in 1755, has impressed the public mind more than any other in modern times. The shocks, one of which exceeded all the others in violence, continued six minutes. The mountains near were shaken to their foundations, and everywhere split and rent. No part of the city was seriously injured which was built on the limestone or basaltic formations; but the shocks were most violent and disastrous in the tertiary and blue clay on which the ruined portion of the city stood.

The sea-wave put in motion by this earthquake exceeded in volume all others of which we have a record, except the one already noticed, which traversed the Pacific Ocean in 1868. It was observed, during this convulsion that the sea retired from the shore before the great wave rolled in.

It was Darwin who first suggested that waves first draw the waters from the shore on which they are advancing to break. He calls attention to the familiar fact that waves thrown up by the paddles of a steamer, as they approach the shore, are always preceded by a receding of the water. An under-draught seems to first suck the water back, and such actually is the fact. Now, in the sea-wave raised by the earthquake, what takes place? We have remarked that an earthquake is a vibration of the earth's elastic crust, and moves with tremendous velocity. When it occurs beneath the sea, or when the undulations reach the surface beneath the sea, the motion is communicated to the water, which it elevates in a wave. Simultaneously with this lifting of the water, an under-draught toward that point takes place. Were it not so, the elevation of the wave could not be sustained. Directly the great wave moves from the area of disturbance at the rate before stated, of 400 miles an hour, or about 6½ miles in a minute, in the deep ocean. It is described by Mallet as "a low, broad swell of the sea. It might pass beneath the vessel unobserved." Approaching the shore, the front becomes elevated. The under-draught has continually preceded it, and has withdrawn the water from the shore, so that vessels at anchor are frequently grounded, and the wave seems to stand upon the bottom like a gigantic wall. At Arica it was unbroken by a ripple, and "shone in the sun like burnished silver."

A notion prevails that earthquakes are always preceded by unusual conditions of the atmosphere, but careful observations have shown that they occur during all kinds of weather. The Lisbon earthquake, which took place in the morning of the 1st of November, was preceded by a "period of clear autumnal weather," but the morning was calm, foggy, and warm. At Arica, as we have learned, the sky was serene and the atmosphere tranquil. Some of the greatest convulsions have been preceded by a close, hazy sky. Sir Charles Lyell observes that "irregularities in the seasons frequently precede and follow shocks. Sudden gusts of wind interrupted by dead calms, violent rains at unusual seasons, or in countries where they seldom occur, are phenomena often attending earthquakes."

The number of important earthquakes up to the year 1881, of which we have a reliable account, is, according to Prof. Ansted, 7,000. So meagre are early records that only 787 of these are spoken of previous to the year 1500. There is a catalogue of 3,340 which occurred from 1800 to 1850, or one in about five days. The means of detecting and recording shocks are now so perfected, that, when applied in all parts of the globe, they will, doubtless, fully justify our statement that in no instant of time is the earth's crust free from vibrations. The seismograph is an instrument for the "automatic registration of earthquake-shocks."

Earthquakes have been defined to be a "travelling zone of vibration." The movement is in every direction from the area of disturbance, and the velocity depends on the substance and structure of the material through which it is transmitted. In New Zealand, in 1848, people on the shore witnessed the disastrous progress of the earthquake along the mountains before they felt the shocks. At Messina, during the

Calabrian earthquake, the terrified inhabitants saw villas overthrown upon the coast by shocks which they had not felt, but which in a moment laid in ruins a portion of their own city. The velocity with which the vibrations travel has been a subject of careful investigation. The Lisbon earthquake moved about 20 miles in a minute; that which occurred in 1819, in the delta of the Indus, appears to have moved at the rate of 53 miles in a minute, or nearly 5,000 feet in a second. Other observations show that the movement may be from 1,000 to 5,000 feet per second. It has been ascertained that in blasting rocks the vibrations move in a second from 1,000 to 1,700 feet. The sound-waves move more rapidly, and, for this reason, shocks are usually preceded by subterranean rumbling. The velocity of sound through uniform strata is ascertained to be from 8,000 to 10,000 feet in a second. Tyndall found that sound-waves moved through burnt clay nearly ten times more rapidly than through air at a temperature of 32° Fahr. From this the phenomena of earthquake movement might occur in the following order : Supposing the centre of the disturbance to be beneath the ocean, as at Lisbon, an observer on the shore might expect to experience —

1. The underground rumble, moving at the rate of 8,000 to 10,000 feet per second.
2. The shock, moving from 1,000 to 5,000 feet per second.
3. The sea-wave, moving about 528 feet per second.
4. Sound, through the air moving at the rate of 1,090 feet per second. It should be noted, however, that the velocity of the sea-wave depends on *depth* of water.

The vibrations of an earthquake, it is evident, differ in no respect from those produced by other causes, excepting in intensity. The jar arising from a discharge of artillery, by a carriage rolling over pavements, or slamming of heavy doors, puts in motion a series of moving waves just as truly as does the rending of rocks, or an explosion of steam or gas in a fracture thus produced. But, a question arises: what moves when the earthquake is progressing. The phenomena maybe explained thus: Around the source of disturbance the rock is pressed outward in every direction as air is pressed outward around a vibrating bell, forming what is called a zone or shell of compressed rock. The extent of this compression is the width of the earthquake-wave, and depends on the force exerted and the elasticity of the rock. In each zone or shell there is always a point of maximum density—and that is where the energy of compression and the rock's elastic force are equal.

As the wave passes, another zone is formed, and the particles behind return by their elasticity to their former position. From this it is obvious that, as the wave is passing, the individual particles of the rock have first a forward and then a backward motion—a swing or excursion to and fro. The extent of this motion is the amplitude of vibration, and may be very small compared with the breadth of the wave.

Mallet found by computation that, given a certain depth of fissure, and a certain heat of steam, the expansive force would produce a wave of nine inches amplitude at the surface. His observations of the Neapolitan earthquake of 1857 show that the maximum amplitude at the surface was only 2.5 inches. In his elaborate and beautiful volume on the eruption of Vesuvius, in 1872, just published, Mr. Mallet reaffirms a statement previously made by him, that "it is the vibration of the wave itself, i. e., the motion of the wave-particles, that does the mischief, not the transit of the wave from place to place on the surface."

We understand, then, that there is motion of particles as well as a transit-wave; that the "travelling zone or shell of vibration" is a zone or shell of "elastic compression."

Now, while the movement of the transit-wave may be very rapid, that of the particles of matter is surprisingly small. At Lisbon the velocity of the wave was 20 miles a minute, or 1,200 miles an hour. According to Mallet, where the velocity of the transit wave was 1,000 feet per second, the movement of the particles was only 12 feet per second, or eight miles an hour, and he states that three columns of the Temple of Serapis, on the shore of the Bay of Baiæ (*see* frontispiece), a region subject to earthquake-shocks, would be overthrown by a shock "whose wave-particles had an horizontal velocity of 3½ feet per second." The shock which threw human beings 100 feet in the air, at Riobamba, must have had a velocity of 80 feet per second. The theory of Mr. Hopkins, published in 1847, was that the disastrous results of earthquakes were caused by the velocity of the wave of translation, and that theory is probably accepted by many who will distrust the conclusions of Mr. Mallet. But it is obvious to every observer that the enormous velocity of 1,200 miles an hour is not communicated to objects on the surface as the wave passes. They are rarely thrown to any considerable distance. Buildings are overthrown, but they fall where they stood.

We have already remarked that objects standing directly on the uniform strata are seldom injured by earthquake-shocks. Such was the case, as we have seen, with that portion of Lisbon which was built on the limestone and basalt. But where the surface, perhaps hundreds of feet deep, is of loose unelastic material, the transit-wave, *with its vibrations,* in passing through it, becomes broken into oscillations, its force is dissipated and motion reduced, but the vibratory swing which it communicates is sufficient to fissure the earth's surface and strew it with ruins.

On the coast of Dublin Bay, Mallet exploded gunpowder buried several feet beneath the surface, in the sand, and ascertained the intensity and velocity of the shock by a delicate seismometer. Other experiments gave the rate and intensity of movement in more compact formations with the following results: In sand, 824.9 feet in a second; in divided granite, 1,306.4 feet; in compact granite, 1,664.6 feet. It is found by observation that objects, as walls and chimneys, fall backward or forward, but generally in a line with the direction in which the wave travels, while fractures of walls occur in a line transverse to the direction of the wave.

The filling of the fissure with water, and its conversion into superheated steam, may have produced the subsequent shocks. By calculation, the same author shows the enormous pressure and rending power of steam if admitted without limit into such a fissure. "If the temperature increase 1° Fahr. for 60 feet depth, then, at the focal centre of the fissure, the temperature would be 883.4° Fahr., and the pressure on the walls of the cavity not less than 640,528,000,000 tons. But the pressure would be vastly increased if the temperature be near that of melted rock." That this may be the case is rendered probable from recent investigations of Mallet, by which he shows that the heat which melts the great lava-beds, and fills cavities in the earth's crust with steam and gases, may not arise directly from the earth's central heat, but from the crushing of strata as it contracts and settles upon the cooling interior.

By a series of experiments and observations made by Mr. Mallet, it is shown that the "annual loss of heat into space of our globe at present is equal to that which would liquefy, at 32° Fahr., about 777 cubic miles of ice; and this is the measuring unit for the amount of contraction of our globe now going on."

The amount of shrinking depends, therefore, on the amount of heat lost—a view long since insisted on by Prof. Dana; and this, according to Mallet, is sufficient to account for all the phenomena. To this cause, then, we refer the never-ending oscillations of the earth's cooled exterior, and the enormous lateral strain by which it is bent and fractured, and its broken ridges made to grind and crush with terrific vibrations.

In many areas the earthquake energies of former times have been long at rest, but, according to Sir Charles Lyell, the total energy may not have diminished.

He finds evidence of convulsions as great and obvious in recent as in earlier time. Mallet, however, remarks that "seismic energy may be considered as possibly constant during historic time, but is probably a decaying energy viewed in reference to much longer periods."

Everywhere we see, in exposed portions, crevices open or filled—ejections of trap and basalt; and wall-like dikes stand out upon the slopes of mountains. These are legible and significant chapters in the earth's dynamic history.

Do earthquakes occur with any order or system, so that their coming may be foretold?

Prof. Palmieri, in his observatory on Mount Vesuvius, is able, says George Forbes, "to predict eruptions." "This is a small eruption," remarked the professor, "but there is going to be a greater one; it may be a year hence, but it will come." "In almost exactly a year," continues Mr. Forbes, "the great eruption did come."

From Mallet's catalogue of European earthquakes it appears that, during 15½ centuries, 1,157 took place during the winter, against 875 in the summer months.

Although science has cleared up some of the mystery which hung over earthquakes in less enlightened times, it has not divested them of their sublimity and terrible reality.

Their work of destruction is done in a moment. The great battles of the world have scarcely been so destructive of human life.

We read that 250,000 persons perished during the earthquake at Antioch in 526. At Lisbon 60,000 people were destroyed. During one of the Calabrian earthquakes 35,000; and during the one at Arequipa in 1868. 40,000 persons perished. Pestilence, famine, and fire, add to the fatality. Visitations so severe and disastrous permanently affect the inhabitants of earthquake regions. Their minds lose their calm equipoise — they become nervous, and the first considerable shock sends them to the street or cathedral for safety.

Humboldt remarks that, when "we feel the ground move beneath us, our deceptive faith in the repose of Nature vanishes, and we feel ourselves transferred into a realm of unknown and destructive forces. Every sound, the faintest motion of the air, arrests attention. To man, the earthquake conveys the idea of unlimited danger." And Von Tschudi adds his testimony, that "no familiarity with earthquakes can blunt this feeling of insecurity. The traveller from the north of Europe waits with impatience to feel the movement of the earth, and with his own ear to listen to the subterranean sounds, but, soon as his wish is gratified, he is terror-stricken, and is prompted to seek safety in flight." Thus it is that physical phenomena aid in moulding the mental and moral character of a people. The earthquake records itself, not only on the inorganic world, but in man's spiritual nature.

Earthquakes and other seismic movements[4]

We are accustomed to think of the land of the earth as something solid and fixed; and, as a testimonial of this impression, the Latin phrase *terra firma,* firm land or solid ground, has been naturalized in the languages of nearly all civilized peoples. On the other hand, we speak of water as unstable. But the geological history of the earth and the more careful observations of modern times have taught us that these ideas do not correctly represent the qualities of the land-masses and water-masses of the globe as compared with one another. The ancient shore-marks on the continents and the phenomena of elevation and subsidence that have been observed in historic times, confirming their evidence, show that the land and the ocean are continually changing their level as to one another; and it has further been made evident,

---

[4] By John Milne

by experiment, as well as by *a priori* reasoning, that it is not the ocean that changes, but the land which undergoes alternate movements of elevation and depression. An earthquake-shock is a phenomenon well adapted to destroy the faith of any person who feels one in the fixedness of the earth; and such, by the evidence, is the effect for the time on all who experience these shocks. Even the light pulsations which sometimes pass over parts of the United States occasion panic and excite a momentary impression that everything is falling over or sinking away; while the more violent shocks that are felt in earthquake-infested countries produce indescribable terror; and such catastrophes as those historical earthquakes of Lisbon and Caracas, and the more recent ones of Ischia and the Strait of Sunda amount to a demonstration that the reason for such terrors is real, and that the continents also cannot escape the general law of change and perishability.

Earth-movements—the name by which these phenomena may be most conveniently described are various, and comprise, so far as they are now considered, earthquakes, or sudden violent movements of the ground; earth-tremors, or minute movements which usually escape attention by the smallness of their amplitude; earth pulsations, or movements which are overlooked on account of the length of their period; and earth oscillations or movements of long period and large amplitude—like the shifting of levels of land-masses—which attract attention from their geological importance. Some of these movements have only recently begun to attract attention. They are all intimately associated in their occurrence and their origin.

The study of earthquakes is of interest to the geologist in many ways. As they seem to be connected with volcanic action, the study of them may help to throw light on that, and *vice versa.* As an earthquake-wave travels along from strata to strata, the study of its reflections and changes in transit may lead to the discovery of peculiarities in rocky structure, of which we should otherwise have no accurate knowledge. It may teach us something about the transmission of disturbances in elastic media, about the earth's magnetism, the electric currents of the earth, and other kindred problems. It is of interest to the meteorologist to know the connections which probably exist between earthquakes and the fluctuations of the barometer, the changes of the thermometer, and the quantity of rainfall. In a practical point, we may ask ourselves what are the effects of earthquakes upon buildings, and how, in earthquake-shaken countries, the buildings are to be made to withstand them.

A typical earthquake consists of a series of small tremors succeeded by a shock, or of a series of shocks separated by more or less irregular—both in period and in amplitude—vibrations of the ground. Man can take but little account of these movements, for they come upon him by surprise, and, by the time he is ready to begin to observe, they are over. Hence we must have recourse to instruments. It is easy enough to construct an instrument that shall move at the time of an earthquake, and leave a record of its motion a—*seismoscope;* but an instrument that shall record the period, extent, and direction of each of the vibrations constituting the earthquake a *seismometer* or *seismograph* is a more complicated affair.

Fig.1

The earliest seismoscope of which we find any historical record is that of the Chinese Chôko, which was invented in A. D. 136. According to the historical account given of it, it consisted of a spherically formed copper vessel (Fig. 1), eight feet in diameter. "Its outer part," the account says, "is ornamented by the figures of different kinds of birds and animals, and old, peculiar-looking letters. In the inner part of this instrument is a column so suspended that it can move in eight directions. Also, in the inside of the bottle, there is an arrangement by which some record of an earthquake is made according to the movement of the pillar. On the outside of the bottle there are eight dragon-heads, each of which holds a ball in its mouth. Underneath these heads there are eight frogs so placed that they appear to watch the dragon's face, so that they are ready to receive the ball if it should be dropped. All the arrangements which cause the pillar to knock the ball out of the dragon's mouth are well hidden in the bottle. When an earthquake occurs, and the bottle is shaken, the dragon instantly drops the ball and the frog which receives it vibrates vigorously. Any one watching this instrument can easily observe earth- quakes. With this arrangement, although one dragon may drop a ball, it is not necessary for the other seven dragons to drop their balls unless the movement has been in all directions: thus we can easily tell the direction of an earthquake. Once upon a time a dragon dropped its ball without any earthquake being observed, and the people therefore thought the instrument of no use, but after two or three days a notice came, saying that an earthquake' had taken place at Rdsei. Hearing of this, those who doubted the use of this instrument began to believe in it again.

After this ingenious instrument had been invented by Choko, the Chinese Government wisely appointed a secretary to make observations on earthquakes." This, the most ancient of the whole class, is closely resembled by some of the instruments of modern times.

The Japanese have an instrument consisting of a magnet holding up a nail, which, when shaken off, starts the train of an alarum, but this does not seem to have ever acted with success. Other seismoscopes depend upon the overthrow of a round column of wood or metal, the projection of balls which are connected with electric circuits, or the disturbance of liquids. Some seismographs depend upon the motions of a pendulum, which may be made to show whether the direction of the shock has been constant or variable, and the maximum extent of its motion in various directions. Other instruments are formed by various adjustments of movable bodies, or with springs and adaptations of clock-work. For a complete seismograph we require three distinct sets of apparatus an apparatus to record horizontal motion, one to record vertical motion, and one to record time. These principles are all embodied in the Gray and Milne seismograph, which is now in use in Japan. In this apparatus (Fig. 2) two mutually rectangular components of the horizontal motion of the earth are recorded on a sheet of smoked paper wound round a drum, D, kept continuously in. motion by clock-work, W, by means of two conical pendulum seismographs, C. The vertical motion is recorded on the same sheet of paper by means of a compensated-spring seismograph, S. L. M. B. The time of occurrence of an earthquake is determined by causing the circuit of two electro-magnets to be closed by the shaking. One of these magnets relieves a mechanism, forming part of a time-keeper, which causes the dial of the time-piece to come suddenly forward on the hands and then move back to its original position. The hands are provided with ink-pads, which mark their positions on the dial, thus indicating the hour, minute, and second when the circuit was closed. The second electro-magnet causes a pointer to make a mark on the paper receiving the record of the motion. This mark indicates the part of the earthquake at which the circuit was closed. The duration of the earthquake is estimated from the length of the record on the smoked paper and the rate of motion of the drum. The nature and period of the different movements are obtained from the curves drawn on the paper.

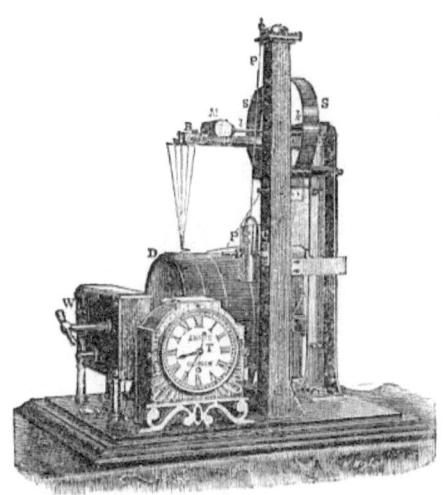

Fig.2

It may be said, as the result of experiences and observations that an ordinary earthquake consists of a number of backward-and-forward motions of the ground following each other in quick succession. Sometimes these commence and die out so gradually that those who have endeavored to time the duration of an earthquake have found it difficult to say when the shock began and when it ended. Sometimes the motions gradually increase to a maximum and then die out as gradually; sometimes the maximum comes suddenly; and at other times during an earthquake the observer's feelings distinctly tell him that there are several maxima. The chief results which investigators have aimed at have been the measurement of the amplitude, period, direction, and duration of the motions; and attention has been given to the velocity with which the disturbance is propagated.

If we were to ask the inhabitants of a town which had been shaken by an earthquake the direction of the motion they had experienced, it is not unlikely that their replies would include all the points of the compass. Many, in consequence of their alarm, have not been able to make accurate observations. Others have been deceived by the motion of the building in which they were situated. Some tell us that the motion was north and south, while others say that it was east and west. A certain number have recognized several motions, and among the rest there will be a few who have felt a wriggling or twisting. Leaving out exceptional cases, the general result obtained from personal observation as to the direction of an earthquake of moderate intensity is extremely indefinite, and the only satisfactory information to be got is that derived from instruments or from the effects of the earthquake as exhibited in shattered buildings and bodies which had been overturned or projected. By the use of seismographs it has been shown that during an earthquake the ground may move in one, two, or several directions, and it is only when a decided shock is experienced that we can determine with any confidence the direction in which the motion has been propagated. The apparently twisting or wriggling motions are supposed to be the result of combinations of linear movements in different directions. It is often difficult, when reading accounts of earthquakes, to determine the length of time a shaking was continuous. Disturbances which succeed one another with sufficient rapidity to cause an almost continual trembling of the ground may be regarded as collectively forming one great seismic effort, which may last a minute, an hour, a day, a week, or even several years. Strictly speaking, they are a series of separate earthquakes, the vibrations of which more or less overlap. Whenever a large earthquake occurs, it is generally succeeded by a considerable number of smaller shocks. Disturbances of this character are compared by Mallet to "an occasional cannonade during a continuous but irregular rattle of musketry." Continuous motions perceptible to our senses without the aid of instruments usually last from thirty seconds to about two or three minutes. The principal vibrations or shocks of the disturbance occur at unequal intervals; and in the periods of vibration there are irregularities in any given earthquake, and different earthquakes differ from one another. The extent of the movement is much less than the feelings of one experiencing a shock would lead him to estimate it. It is usually within the fraction of an inch in either direction. According to Dr.

Wagener, the earth's horizontal motion at the time of a small earthquake is usually only the fraction of a millimetre, and it seldom exceeds three or four millimetres. Mallet believes that the displacement may in some instances be equal to a foot; and M. Abella records a rough observation, in the Philippine Islands, of a motion of the earth to a distance of two metres, when fissures were formed, and seen to open and shut. The velocity of propagation of the wave may vary, even in the same country, between several hundreds and several thousands of feet per second. The same earthquake travels faster across districts near to its origin than it does across districts which are far removed; and, the greater the intensity of the shock, the greater is the velocity.

If we were suddenly placed among the ruins of a large city which had been shattered by an earthquake, it is doubtful whether we should at once recognize any law as to the relative position of the masses of rubbish and the general destruction around. The results of observation have, however, shown that, among the apparently chaotic ruin produced by earthquakes there runs more or less of law governing the position of bodies which have fallen, the direction and position of cracks in walls, and the other effects. Usually, walls of buildings at right angles to the shock will be more likely to be overthrown than those which are parallel to it. It is said that in Caracas every house has its lado securo, or safe side, where the inhabitants place their fragile property. It is the north side, and has been chosen because about two out of three destructive shocks traverse the city from west to east, so that the walls in those sides of the building take them broad- side on. This appears to be the rule in destructive earthquakes. But, when a building is subjected to a slight movement, it is assumed that the walls at right angles to the direction of the shock move backward and forward as a whole, and there is little or no tendency for them to be fractured at their weaker parts. The walls, however, which are parallel to the direction of the movement are extended and contracted along their length, and may consequently be expected to give way over the door- and window-openings. The results of the examination of more than three hundred foreign-built brick houses in Tokio, Japan, all similar in their construction, are typically illustrated in Fig. 3. They show that in the upper windows nearly all the cracks ran from the springing of the arches, which formed an angle with the abutment. In the lower arches, which curved into the abutments, not a single crack was observed at the spring-way. The cracks in those arches were near the crown, where beams projected to carry the balcony; and in many instances they proceeded from such beams, even if there were no arches beneath. The houses which were most cracked were in the streets running parallel to the direction in which the greater number and most powerful set of shocks cross the city. From the fact that cracks once made in a building did not appear to extend under the repetition of shocks similar to the one that produced them, it has been inferred that buildings thus cracked acquire a degree of flexibility, and that, by providing cracks or joints between the parts of buildings which have different periods of vibration, some of the strain might be taken off from them, and they might be made more stable. In stone-work, the cracks have been observed generally to run through the mortar-joints; in brick-work, through either bricks or mortar, often preferring the bricks.

As fractures in walls seem most likely to take place above openings like doors or windows, it follows that where architecture demands that openings should be placed one above another in heavy walls, there will be lines of weakness running through the openings. As arches are only intended to resist vertical thrusts, special construction must be adopted to make them strong enough to resist horizontal pulls. This might be given by inserting iron girders or wooden lintels in the arches. Mr. Perry, of Tokio, has suggested a plan of building so that the openings of each tier would occupy alternate positions. Such a line is shown in Fig. 4, where the dotted lines run through openings representing the direction of the lines of weakness. If we compare this with Fig. 5, we shall see that in the case of a horizontal movement, *a b,* or a vertical movement, *c d,* fractures would more probably occur in a house built like Fig. 5 than in one built like Fig. 4. If, how-ever, these two buildings were shaken by a shock which had an angle of emergence of about 45°, in the direction of *e f,* the effects might be reversed. Fractures following a vertical line of weakness are shown in the accompanying drawing (Fig. 6) of the church of St. Augustin, at Manila, shattered by the earthquake of 1880.

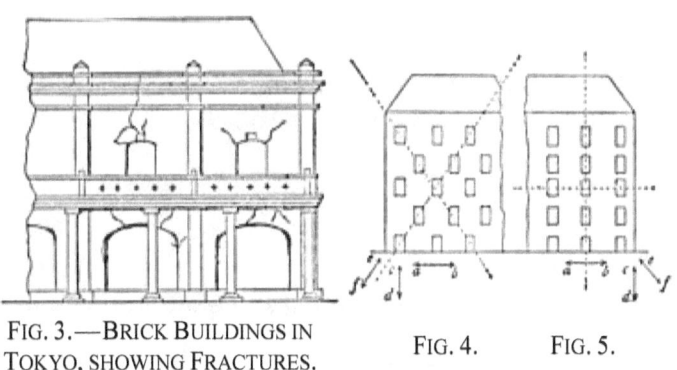

FIG. 3.—BRICK BUILDINGS IN TOKYO, SHOWING FRACTURES.

FIG. 4.   FIG. 5.

FIG. 6.—CHURCH OF ST. AUGUSTIN, MANILA. EARTHQUAKES OF JULY 18-20, 1880.

When an earthquake shock enters and proceeds along a line of buildings, the last building in the row will, of course, suffer the most, and will exhibit the greatest tendency to fly away from its neighbors. If the house stands on the edge of a canal, or cliff, this tendency is increased by the similar motion of the escarpment. The fate of an end-building thus stricken is shown in Fig. 7, which is taken from the photograph of a house that was shattered in 1868 at San Francisco. Houses may also be rocked on their foundations, or even quite overturned, as appears to have happened to the stud-mill at Hayward, California (Fig. 8).

FIG. 7.—WEBBER HOUSE, SAN FRANCISCO, OCTOBER 21, 1868.

Fig. 8.—Stud Mill at Haywards, California, October 21, 1868.

In any building which may be affected by an earthquake, we have to consider the vibration of a number of parts, the periods of which, if they were independent of each other, would be different. On account of this difference in period, while one portion of a building is endeavoring to move toward the right, another is pulling toward the left, and either the bonds which join them or the parts themselves will be strained or broken. This was illustrated by many of the chimneys in the houses at Yokohama, which, in the earthquake of February 20, 1880, were shorn off just above the roof. Since then, builders have learned to let chimneys pass freely through the roof without coming in contact with any of the main timbers.

In trying to make structures earthquake-proof, we may build our house weak and flexible, so that the shock shall pass over it as the wind over a reed, or we may attempt to make it stronger than the shock. The native Japanese houses, with their flexible framing, are built on the former plan; some of the European houses essay the latter. In Italy the houses are left to take their chances. In South America, where much exposed to earthquakes, they are built of only one story, or of bamboo and ropes, similarly to the Japanese plan. One of the safest houses for an earthquake country would probably be a one-storied, strongly framed timber house, with a light, flattish roof, made of shingles or sheet-iron, the whole resting on a quantity of small cast-iron balls carried on flat plates bedded in the foundations. The chimneys might be made of sheet-iron, carried through holes free of the roof. The ornamentation ought to be of light materials. The nature of the ground on which the house is built does not always appear to be in itself a matter of prime moment. Its relations with other foundations are more important. In some places solid strata, in others soft strata, appear to afford the more favorable situations; and the superiority of either probably depends on a variety of local circumstances. Places near the junction of the two kinds of formations are the worst. The progress of the wave may be interrupted by the interposition of a mountain-range or a hill, in which case we have behind the barrier the phenomenon called an earthquake-shadow; it may be cut off by a deep ditch, as a canal; and in certain parts of South America there appear to exist tracts of ground which are practically exempt from the shocks, while the whole country around is violently shaken. It would seem as if the shock passed beneath such a district as water passes beneath a bridge; and for this reason such tracts have been christened "bridges." In the Syrian earthquake of 1837, neighboring villages, and even neighboring houses, suffered differently. In one case a house was entirely destroyed, while in the next house nothing was felt. In

Japan, at a place called Choshi, about fifty-five miles east of the capital, earthquakes are seldom felt, although the surrounding districts may be severely shaken. At this place a large basaltic boss rises in the midst of alluvial strata. The immunity of the district from earthquakes has probably given rise to the myth of the Kanam rock, which is a stone supposed to rest upon the head of a monstrous cat-fish, whose writhings cause the shakings so often felt.

Possibly something may be done in arranging the surroundings of buildings to ward off the destructive effects of earthquakes. The Temple of Diana, at Ephesus, was built on the edge of a marsh for this object. Pliny says that the Capitol of Rome was saved by the Catacombs. Elisee Reclus says that the Romans and Hellenes found out that caverns, wells, and quarries retarded the disturbance of the earth, and protected edifices in their neighborhood. The Tower of Capua was saved by its numerous wells. Vivenzis asserts that in building the Capitol the Romans sank wells to weaken the effects of terrestrial oscillations; and Humboldt relates the same of the inhabitants of San

Domingo. Quito is said to receive protection from the numerous canons in the neighborhood, while Lactacunga, fifteen miles distant, has often been destroyed. Similarly, it is extremely probable that many portions of Tokio have from time to time been protected more or less from the severe shocks of earthquakes by the numerous moats and deep canals which intersect the city.

Various causes have been assigned for the production of earth- quakes, and, although they may all singly or in combination contribute to the effect, we must conclude, after considering the whole subject, that the primary cause is endogenous to our earth, and that exogenous causes, like the attraction of the sun and moon, and barometric fluctuations, play but a small part in the actual production of the phenomena, their greatest effect being to cause a slight preponderance in the number of earthquakes at particular seasons. The majority of earthquakes are due to explosive efforts at volcanic foci. The greater number of these explosions take place beneath the sea, and are probably due to the admission of water through fissures to the heated rocks beneath. A smaller number of earthquakes originate at actual volcanoes. Some earthquakes are produced by the sudden fracture of rocky-strata or the production of faults. This may be attributable to stresses brought about by elevatory pressure. Lastly, we have earthquakes due to the collapse of underground excavations; and these may have been produced by evisceration caused by volcanic eruptions, by the washing away or solution of the earth by chemically charged waters or hot springs, or by other causes.

Considerable attention has been drawn lately toward the study of small vibratory motions of the ground which, to the unaided senses, are usually passed by without recognition. They are called earth-tremors, and were only discovered when difficulties caused by them were encountered in the adjustment of extremely delicate astronomical and other instruments.

These movements have been most carefully studied in Italy by Father Bertelli, of Florence; le Conte Malvasia, at Bologna; M. di Rossi, at Rome; and le Baron Puet, at Nice.

Delicate instruments have been devised for detecting and recording them, the most important of which is the *normal tromometer* of Bertelli and Rossi. It consists of a pendulum (Fig. 9) one and a half metre long, carrying, by means of a very fine wire, a weight of one hundred grammes. To the base of the bob a vertical style is attached, and the whole is inclosed in a tube, terminated at its base by a glass prism of such a form that, when looked through horizontally, the motion of the stile can be seen in all azimuths. In front of this prism a microscope is placed. Inside the microscope is a micromatic scale, so arranged that it can be turned to coincide with the apparent direction of oscillation of the point of the stile. In this way not only can the amplitude of the motion of the stile be measured, but also its azimuth. The extent of vertical motion is measured by the up-and-down motion of the stile due to the elasticity of the supporting wire.

]

Another instrument, the *microseismograph* of Professor Rossi, gives automatic records of slight motions. It consists of four pendulums, each about three feet long, suspended so that they form the corners of a square platform. In the center of this platform a fifth but rather longer pendulum is suspended. The four pendulums are each connected just above their bobs to the central pendulum with loose silk threads. Fixed to the center of each of these threads, and held vertically by a light spring, is a needle, so adjusted that each thread is de- pressed to form an obtuse angle of about 155. These needles form the terminals of an electric circuit, the other termination of which is a small cup of mercury placed just below the lower end of the needle. By a horizontal swing of one of the pendulums this arrangement causes the needle to move vertically, but with a slightly multiplied amplitude. By this motion the needle comes in contact with the mercury, and an electro-magnet with a lever and pencil is caused to make a mark on a band moved by clock-work. The five pendulums being of different lengths, the apparatus is adapted to respond to seismic waves of different velocities.

Professor Rossi's microphone consists of a metallic swing arranged like the beam of a balance. By means of a movable weight at one end of the beam, this is so adjusted that it falls down until it comes in contact with a metallic stop. The beam and the stop form two poles of an electric circuit, in which is a telephone. The slightest motion in a vertical direction causes a fluctuation in the current passing between the stop and the beam, and announces itself in the telephone.

By observations made with instruments like these, it has been shown that the soil of Italy is in incessant movement, with periods of excessive activity, called seismic storms that usually last about ten days. The storms are separated by periods of relative calm. They are more regular in winter, and exhibit sharp maximums in spring and autumn. In the midst of such a period or at its end there is usually an earth- quake. They have been observed to be generally related to barometric depressions.

Earth-pulsations are slow but large undulations that appear to travel over or disturb the surface of the globe. They are made manifest through variations in the movement of pendulums, changes in the position of the bubbles of levels, eccentricities in the behavior of clocks, the swinging of chandeliers in churches, unusual disturbances in bodies of water, and even of water in tubs, irregularities in the flow of springs, and other phenomena, the occurrence of which, or the peculiar manner of it, while it is consistent with the hypothesis of such movement, cannot be accounted for on any other probable supposition.